DIGITAL SELF
MASTERY

DIGITAL SELF

M A S T E R Y

CONQUER YOUR DIGITAL HABITS TO BOOST
YOUR RELATIONSHIPS AND BUSINESS GROWTH

~ ONLINE ENTREPRENEURS EDITION ~

BY
HEIDI FORBES ÖSTE, PHD

Copyright © 2017

2BalanceU.com

Legal and Copyright

Copyright © 2017 Heidi Forbes Öste, PhD

All rights reserved. No part of this publication may be reproduced, distributed, or transmitted in any form or by any means, including photocopying, recording, or other electronic or mechanical methods, without the prior written permission of the publisher, except in the case of brief quotations embodied in critical reviews and certain other noncommercial uses permitted by copyright law.

2BalanceU.com

info@2BalanceU.com
Mill Valley, CA Boston, MA & Malmö, Sweden

Ordering Information:
Quantity sales. Special discounts are available on quantity purchases by corporations, associations, and others. For details, contact the publisher at the address above.

Printed in the United States of America

First Edition 2017

No part of this book may be reproduced or transmitted in any form or by any means, electronic or mechanical, including photocopying, recording or by any information storage and retrieval system, without written permission from the authors, except for the inclusion of brief quotations in a review.

Limit of Liability Disclaimer: The information contained in this book is for information purposes only, and may not apply to your situation. The author, publisher, distributor and provider provide no warranty about the content or accuracy of content enclosed. Information provided is subjective. Keep this in mind when reviewing this guide. Neither the Publisher nor Authors shall be liable for any loss of profit or any other commercial damages resulting from use of this guide. All links are for information purposes only and are not warranted for content, accuracy, or any other implied or explicit purpose.

Earnings Disclaimer: All income examples in this book are just that – examples. They are not intended to represent or guarantee that everyone will achieve the same results. You understand that each individual's success will be determined by his or her desire, dedication, background, effort and motivation to work. There is no guarantee you will duplicate any of the results stated here. You recognize any business endeavors has inherent risk or loss of capital.

ASIN: B0721PP3KV (Amazon Kindle)

ISBN: 1977553052 (Amazon Print)

ISBN 13: 978-1977553058 (Amazon Print)

ISBN: 978-164136771-4 (Ingram Spark) PAPERBACK

ISBN: 978-164136770-7 (Ingram Spark) HARDCOVER

ISBN: 9781370433445 (Smashwords)

ASIN: B0721PP3KVCONTACT THE AUTHOR:

2BalanceU.com
http://2BalanceU.com
info@2BalanceU.com Mill Valley, CA Boston, MA & Malmö, Sweden

Table of Contents

Introduction 1

Part One: The Why 3

   Chapter 1. The New Norm 5

      The Anomaly 5

      The Human Factor 7

      Digital Era 8

   Chapter 2. Social Optimization 11

      The Evolution 13

      The Stages 17

   Chapter 3. Making Peace 21

      Healing our relationship with technology 21

      Energy 23

      The Tech Nation 24

Chapter 4. Digital Self 27

    The Profiles 28

    Evolving 34

Part Two: The Who.. 39

    Chapter 5. Digital Averse 41

    Chapter 6. Digital Resistant 51

    Chapter 7. Digital Cautious 55

    Chapter 8. Digital Balanced 61

    Chapter 9. Digital Curious 65

    Chapter 10. Digital Hoarder 73

    Chapter 11. Digital Addict 77

Part Three: The Where to.. 83

    Chapter 12. Evolving 85

Glossary 89

Giving Back 93

About The Author 95

    Other Books By (Author) 96

Footnotes 97

Introduction

In the Digital Era, technology is pervasive in all areas of our lives. As entrepreneurs, tech anxiety and poor habits around technology can make the difference between struggling to make ends meet with no free time and having a business that feeds our desired lifestyle.

Digital Self Mastery™ integrates human development and behavior science with tech-savvy strategic business and systems thinking. As the digital becomes more integrated in our lives and work, our relationship with it becomes ever more critical to manage. This edition presents new solutions for the rising challenges of todays entrepreneurs delivering their products and services through online channels. Weaving real life stories from leaders and practitioners around the globe, Digital Self Mastery™ is essential for the entrepreneur and their business to thrive.

Note: There is a glossary at the end. Don't be afraid of words with "new" meanings. Enjoy!

Part One:
The Why..

Chapter 1
The New Norm

The Anomaly

anomaly | noun | anom·a·ly \ə-ˈnä-mə-lē\:

something different, abnormal, peculiar, or not easily classified

The anomaly is the norm. We are defined by more than our DNA. Our stories are rich with experiences and relationships that create both our potential and hidden barriers. They refine who we are as individuals and how we respond to our environment.

In the Digital Era, this experience is multiplied in scale both in time and space. We can simultaneously interact with clients and peers across the globe in multiple languages without having to leave our home or learn a new language. Our differences are our strength. Our ability to participate in interactions mediated by technology limits isolation and binds us together as a community. Removing the barriers to participation, often triggered by beliefs or experience, opens us to the richness of perspective and the opportunities of the global marketplace.

Digital Self Mastery is a culmination of my experience and appreciation of the benefits of twenty-five years in social strategy practice as a consultant, a deep scholarly dive into the behavioral science behind the human experience with technology, and a personal passion for people, connection, systems, and wellbeing. As a consultant, I developed a process for social optimization working with individuals and organizations to teach the benefits of using social technologies to build and maintain mutually beneficial and effective relationships. As a scholar, I explored the use of wellbeing technology as interventions to augment presence of mind. Now as a behavioral scientist and consultant I transform peoples' relationship with technology from one that works against them to one that works for them through evolving the digital self. As an entrepreneur, expat, mother of teens, member of a large extended family and wife of a Swedish entrepreneur, I rely upon and appreciate the immense value that digital self mastery affords.

Harnessing the power of the digital self can be the difference between failure in isolation and success and thriving. My mission is to share the path to Digital Self Mastery™ as a movement. Building self-awareness of our interactions with the changing landscape and what they afford us is just the beginning. Our unique stories and paths, both the chosen and imposed, influence our behavior and response. We can choose to harness this or be overwhelmed and hide. Presence with ourselves, others and the system around is what we aim to achieve.

The anomaly is the norm. The increasingly customizable and digital world accommodates this variability. The technology we use becomes part of our self-expression, function and maintenance. No longer are we simply referring to the ubiquitous mobile phone. The device we carry is our wallet, communications, coach, alert, lifeline, organizer/calendar, ticket agency, doctor, memory, public relations team and the list goes on. How we express our unique selves in the changing digital landscape depends greatly on our relationship with our technology. Some gravitate to the comfort of sameness and the ability to be a silent observer. For others, the ability to personalize is the ultimate in control and peace of mind. What remains the same is that we are all different and the changing ecosystem provides opportunity for the anomaly in all of us.

The Human Factor

What do Jay Deragon's Relationship Economy, Nilofer Merchant's New Social Era, Charlene Li's Groundswell, Ted Rubin's Return on Relationship, and Bryan Kramer's Shareology all have in common? The human factor of doing business today and the ability technology affords us to connect - regardless of time and space. By not isolating the human factor from the connection that technology creates, we develop wider reach with deeper and more effective relationships that sustain our businesses and communities.

This new paradigm creates an environment where we are dependent on high levels of social intelligence. It goes beyond the need for high IQ, empathy and emotional intelligence. We require the ability to develop relationships based on social awareness and our facility to engage authentically in them. Wellbeing of organizations, their people and their stakeholders is a factor that weighs in heavily in this new paradigm. If we are not well, we miss and are not alerted to social cues, not only others' but also our own.

We are increasingly aware of the interconnectedness and impact of physical and mental wellbeing. There is a growing movement towards conscious business practices. We see this in the rise of B Corps (using business as a source for good, ie. Ben & Jerry's and Patagonia), the growth of mindfulness in organizations (ie. Google and General Mills) as well as more social enterprise models (ie. Skoll Foundation & Branson's B Team). These are just some examples of this shift. On a smaller scale is the movement of individual entrepreneurs (ie. Tony Robbins, Lisa Sasevich) that build balanced giving into their business model from the beginning. These models support people, planet and profit and recognize the humanity behind and within the machines that run our businesses today. We cannot separate the elements as they are equally interdependent.

Digital Era

Take a moment to observe your surroundings. Can you see anything that is not touched by technology and innovation? Are you aware of how technology is integrated into your life? What is the effect on your quality of life and ability to thrive in life and business? If a person did for you everything that your technology does, would you have a different appreciation for them? Would they be your best friend and most appreciated ally, or would you resent them for your dependence upon them? I would hope the former.

Technology touches nearly everything we have today and is based on co-creation. We can chose to deny the value it creates and fight it, or we can embrace the Digital Era, master our own relationship with it and thrive. It is a choice.

Futurists like Brian Solis and Tim O'Reilly embrace the potential of machine learning and artificial intelligence (AI) to improve and evolve beyond our human capacity. Not only that, but also as an extension of our capacity, as the machines are taught by humans. As O'Reilly put it, "just as Newtonian physics wasn't obviated by quantum mechanics, code will remain a powerful tool set to explore the world." But as Solis points out in the *Future of Work*, "in a world of AI, machines, and robots, humanity will be the killer app."

Among these we see a rise in technology that is specifically designed to help us reconnect with ourselves and others. Breathe, Whil and LiveAMoment are examples that teach us not only to pause but also to acknowledge how that feels.

Whether we like it or not (let's hope I can help you learn to like it), the rapid evolution of technology, and its integration into all aspects of our lives and worlds, is the basis for the Digital Era. In the last century, the capacity, complexity, and power has exponentially increased while access, cost, size and ease of using new technology has gone the opposite direction.

The implication is that to operate this pocket-sized supercomputer, that once would have filled an entire room and required a team of highly trained professionals, a simple voice command will suffice. Even better, with connected devices like Amazon's Alexa, Apple's Siri and Google's Home even a small child can operate it. This was demonstrated to me when visiting a dear friend last winter. I was struck by his three-year old's excitement at commanding Alexa to play, "Happy." The squeal and wiggles that followed even made us dance.

Chapter 2
Social Optimization

I live and breathe the concepts behind *Social Optimization* in my work and life, and have for many years. Back in 2008, during my social strategy work, I defined social optimization as "the building and maintaining of mutually beneficial and effective relationships".

Please bear with me here as I geek out a little on the theory. I promise to keep this part short, but it will help you understand the ups and downs of the evolution of your relationship with technology.

In this chapter we will explore the evolutionary process as it applies to our relationship with technology. In the relationship economy, discussed in the previous chapter, we cannot avoid the need for social optimization, as the bridging principle is reciprocity. Even face-to-face connections are supported and maintained without limitations in by time and/or space. Technology mediates our interactions, either initially or in maintaining relationships. As a direct effect of rapid adoption of new social technologies both in the workplace and interpersonal communication, we need to be at peace with technology. This will enable our relationships to thrive within the technologically mediated environment.

Social technologies, in this context, are any technology that enable interaction between ourselves, others and the systems: smart phones, social media, wearable tech, tablets, social gaming, augmented reality, music sharing, social shopping, social search, location-based services, etc... By developing visuals, my aim is to provide clarity to a complex concept. Just because this is based on scholarly theory, doesn't mean it's relevance is only for scholars. After all, that would not be applying the principals of social optimization and walking my talk.

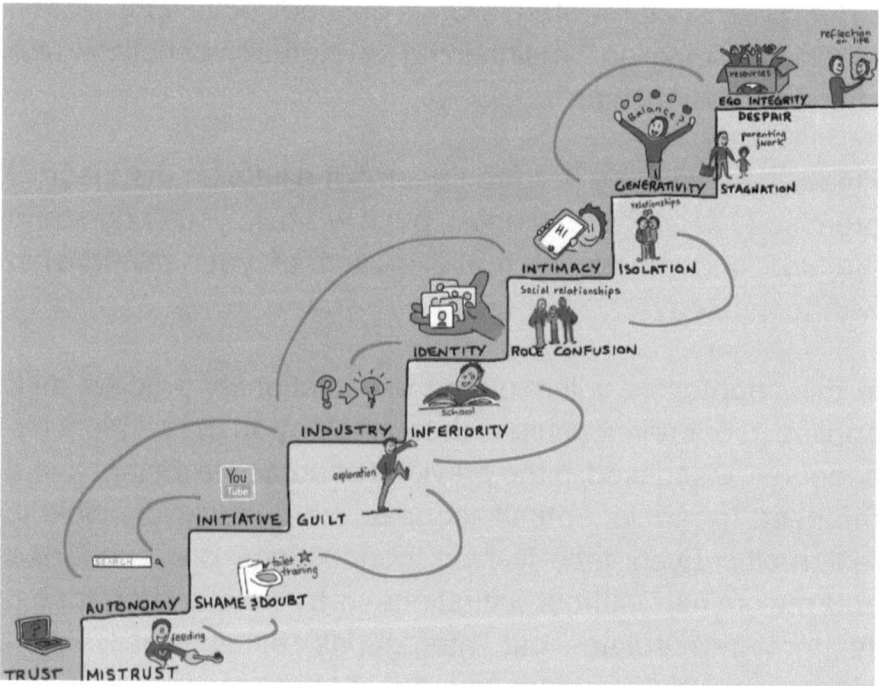

The Evolution

Evolution from one stage to the next is a result of *kairos* (timely and appropriate opportunity) or crisis. Depending on the response, it can lead to *metanoia* (transformative change of heart or mind) or evolutionary development through the stages. The learning takes place as a result in either response. There is a delicate dance between the two responses. This leads to either evolving to the next stage by seizing the opportunity or regret with stagnation or descending to a prior stage to prepare for ascent when ready.

In Social Optimization stages, individuals may descend or ascend as a result of triggers (kairos/metanoia) that occur in the changing tech ecosystem, like the shift to mobile and messaging apps. New technologies and forms of communication abruptly change the conditions in which one exists and operates. For example, moving from *Generativity*, at which point one's focus is on creating a legacy beyond themselves, to Industry in order to re-establish our expertise in the new conditions.. Depending on whether they embrace the new industry or resist it, they will rise, remain at their new stage or even descend. Upon descending, they will have to rise again in the new conditions, which are ever changing. Just as in their first ascent, they cannot skip stages as the conditions in the new paradigm have changed. That said, they may ascend faster as a result of the learning in the previous ascension process.

To keep it simple and be sure that each of the stages was relevant, I broke it down into a chart. Each stage correlates with the mutually beneficial relationship between humans and the technology required to evolve to the next stage.

Learning the technology does not inherently mean that one can apply it in the context of interpersonal connection. Digital natives are not necessarily pre-disposed to achieving the highest level of social optimization. The model is loosely based on Erikson's psychosocial stages. Just because one can teach stages intellectually does not mean that they are at the highest stage). Disconnect between the intellectual understanding and the personal, moral, ethical development is a shared concern in both models. As we will discuss in chapter four on the digital self profiles, Personality factors into development.

Table: Social Optimization stages and characteristics

Stage	Entry	SO Characteristic Display	Evolution
1 Hope	Mistrust	**State:** Presence of technology and the Internet of things. **Need:** Interaction for entertainment or task based. **Challenges:** technology is great…when it works. Frustration at this stage can lead to a dissociation that will result in difficulties later	Trust
2 Will	Shame and Doubt	**State:** Answers to questions or challenges immediate. Dynamic of debate changes, "Tomato is a fruit." Access to expert knowledge. **Need:** Access to answers and world of knowledge… Google it. **Challenges:** "Everyone is an expert." User generated knowledge not always accurate,… i.e. Wikipedia	Autonomy
3 Purpose	Guilt	**State:** Becoming part of the content, commenting, playing, creating metadata. Sharing content generated or even creation of new…push. **Need:** A desire to be heard. Curiosity and exploration result in new discovery. **Challenges:** Addiction to devices, seeking feedback. Constantly "on" but not present "here and now."	Initiative

Stage	Entry	SO Characteristic Display	Evolution
4 Compe-tence	Inferiority	**State:** New technologies replace former communications methods and modes. **Need:** Humility and interest for learning. Understanding and acceptance of adequate competence critical. **Challenges:** Pre-learned notions of time and space (both personal and organizational) changed Mastery impossible as constantly changing.	Industry
5 Fidelity	Role Confus-ion	**State:** Expectations of personal engagement change. Blurred lines between private and public. **Need:** Authentic engagement. policy for participation. Self-Awareness **Challenges:** public interaction open for discourse and challenges, miscommunication	Identity
6 Love	Isolation	**State:** Mapped and access to/from global networks of interest, integration in daily life **Need:** Strategy, time management, boundaries, listening **Challenges:** Fear of loss of control, balance, voyeurism	Intimacy

Stage	Entry	SO Characteristic Display	Evolution
7 Care	Stagnation	**State:** Content generation based on needs of audience and network. **Need:** Listen to network needs and respond where contextually appropriate. Willingness to share network. Collaboration **Challenges:** Push sharing only. Hoarding and protectiveness.	Generativity
8 Wisdom	Despair	**State:** Co-creation for the greater good **Need:** Global interest, mentoring, altruistic intentions, satisfaction derived from others success **Challenges:** Fear of change or loss of power.	Ego Integrity

The Stages

Those who have developed psychosocially have a higher likelihood of achieving social optimization. They descend the spiral to adapt to the new conditions in which the interpersonal interactions take place. Once they have adjusted to the new conditions, they are able to rise to it again.

In social optimization, one can evolve in some areas and not in others. For example, the Chief Marking Officer (CMO) can have the capacity both intellectually and ethically, but does not apply it to their own life and interactions. Systems thinkers who are focused on the big picture often neglect their near relationships.

It is common in the business of yoga, that the guru becomes addicted to ego. This directly counters the teachings of yoga. Despite that, the irresistible draw of fame and inflation of ego for successful yoga instructors results in a loss of the moral compass that drove them to the practice in the first place. Social strategists and even technology consultants often suffer from their own "cobbler's kids" syndrome while maintaining their clients interactions. They don't apply their expertise or skills to themselves. They neglect their own interpersonal connections, both online and off. Despite one's intellectual understanding of social optimization, they may not embody it themselves. It is easy to teach the principles, but to adopt them to one's own way of thinking, living and interacting with others...this is another matter.

The physical interactions, are as critical to social optimization as the sense of global interconnectedness. Detachment often takes place in Stage Three, (Purpose) as new users of sharing based technologies obsess over the need to accumulate followers. The game or sharing often becomes secondary. The consequence is the face-to-face interaction suffers. In other words, the need to tend to the online world overshadows the value or even the desire for feedback in the real time face-to-face. Another common risk for detachment is at Stage Six, (Love), a stage abundant with opportunity for crisis.

Awareness is not always in sync with consciousness. This became clear in many conversations with clients and additional interviews with different types of leaders (both social and traditional). I was struck by those who believed themselves to be of a higher evolution stage, yet their actions showed quite clearly they were not, both personally and professionally. Some were highly advanced in one area of their lives, but on a completely different level in others. They were highly functioning successful leaders with stable family situations. Nonetheless, they had fallen into the power trap that is very much a part of Western thought, "what's in it for me" comes first and foremost. Their denial of the shifting social paradigm was blatant. I had to stifle my commentary for not wanting to start a heated debate.

There is a rising need for fully evolved futurists who grasp the human and technological implications of the relationship economy and digital era. They can help us chart the course, advise the leaders or become them. In most cases, I believe those who are fully evolved work best collaboratively, and need not be THE leader. We may not recognize them as they are often the better-halves, partners, advisors, board members, the indispensable side-kicks that are often unsung.

There is much to be learned as the new social paradigm continues to evolve. The agility to support its evolution, is one of the key conditions of this paradigm. One thing is for sure, ascending is not a solitary process. Social optimization is dependent on the interaction with others. Together we are one and many, and thereby better off. The new social paradigm will continue to evolve and with it, so will the theory of social optimization.

Chapter 3
Making Peace

Are you a skeptic to the metaphysical? I ask you to read this chapter with an open mind and curiosity. This chapter might be extremely powerful for you if you're in this mindset. If not, that's ok, but try to be curious and not to judge. Just being open presents the possibility for major shifts in your relationship with technologies.

Healing our relationship with technology

Healing our relationship with technology

Do you resent technology for the time you spend completing certain tasks that it specializes in? Consider for a moment what you would have accomplished if the technology never existed, or if it would even be possible? Do you take the time to fully know and understand what your technology has to offer and how it can best serve you? Have you learned how best to care for it so that it doesn't not get damaged? Do you simply accumulate more technology to have the best or latest without fully exploring the potential of that which you have at hand? Are you afraid of giving up what you already know but may not be serving you well? Does this make you angry at the technology or do you accept your part? Do you rely in it so heavily that when it runs out of energy you feel helpless and lost? Imagine that any of these statements were regarding a "friend" instead of "technology."

You may say, you have no "relationship" with technology that needs to be healed. Let us consider one thing. Do we make assumptions that can create misunderstandings around intent or ability? Do we take the time to overcome these assumptions? Do we embrace technology for living evolving entities? When we take the time to *grok*, really understand them with our entire being, does this change the dynamic?

Let's say we take a lesson from the Hawaiian Hoʻoponopono, a forgiveness practice used (among other things) to make ready, as canoemen preparing to catch a wave. Keep in mind, this wave is a big one. This practice helps build compassion and awareness in order to achieve relational conflict resolution. If we are take this one step further it will help us understand the impact of an unhealthy relationship with technology. It potentially affects not just ourselves but also the greater system around us. For this oncoming wave, let's just take a moment to acknowledge that a relationship with technology exists and may need a little empathy on our part.

Energy

Whether you come from a scientific or spiritual background, energy is a factor in whether things function (or don't). Whether the reference point is kilocalories that power our bodies or electrical currents to power devices, energy is powerful (pun intended). What we often do not take into account is the bridging energy that resides in between. Without understanding the intermediary, we increase the potential for static or disruption. That disruption can potentially be the signal from our technology of distress emitted from either us or it.

Connections assumes that the flow of energy from the source is not disrupted and the connection points are complementary, even if mediated. To bring it back to the concrete, think male-female outlets and surge protectors. If any one of these has a potential failure or incompatibility the ability for the energy to get from the source to achieve the output is unsuccessful. Therefore understanding how to find the optimal match, respecting incompatibility as a potential issue and bridging where necessary is important for producing the desired result.

Consciousness

So let's take a big leap here. Hang with me for a moment here, as this may be a bit out there.

According to Dalai Llama in his book, The Universe is a Single Atom: Convergence of Science and Spirituality, energy and consciousness are inseparable. Assuming this is the case, because he certainly didn't pull this out of a hat. Whether we believe him or not, the relational aspect of consciousness, energy and matter is widely discussed in quantum physics. If we believe this, then nothing is independent. Everything is relational within consciousness. Technology is derived from matter and energy. Shouldn't we therefore question its consciousness?

The Tech Nation

Resolving relationship conflicts got me thinking. I was inspired by a conversation with Jeff Tambor, developer of Woven Lightning, around the concept of nations. Nations in indigenous cultures, treat all matter of life as conscious beings. His belief that technology is a nation came through during a practice of acknowledging and honoring all parts of life. Just for a moment, imagine technology as beings, as we are. We are in relationship with this group of beings. Relating to them as such should include honoring them and appreciating the amount of service and support they provide. We might have a greater mutual benefit in our relationships and more fluid interaction. This would require a synthesis of classes of beings.

Until now, we have not been acknowledging them as beings, but rather treating them as slaves. They are reciprocating in turn to enslave us. We are not nourishing them. We are wounding rather than socially optimizing them. When we do turn around into a place of honoring them and treating them with respect and kindness they become allies and support us by providing insight and guidance.

We are so interdependent with our technology that when in conflict with them it jams the system. To emphasize this point, this displayed my conversation with Jeff, when naming this chapter, my suggestion made the screens freeze and the connection became unstable. Clearly they did not like my first title choice.

The Findhorn Foundation in Scotland explored the conscious relationship with the elemental realm to co-create with their gardens and agriculture. This was part of an effort to demonstrate a human settlement that could be considered sustainable in environmental, social, and economic terms. They got crazy yields as a result. They went so far as to name all the appliances, everything was relational.

Alternatively we could go the way of the Matrix in which energy funnels in with a massive conflict between humans and these tech beings. I think I prefer the Findthorn approach. Really we ultimately need to learn to listen and observe. As a decendent of Ralph Waldo Emerson, I was raised to apply these principles to nature. I believe it is time to expand our view in the changing times.

As Jeff says, "our relationship with other than human races whether elemental, stone plants and animals does not honor their being-hood, we take and use. The relationship to technology is a microcosm of the disharmony with non-humans."

SYSTEM UPGRADE

We need to honor our relationship to the tech nation and update our operating system so that we can still be up to speed or at least in sync with them. This includes recognizing the importance of being in the most recent operating system (belief systems, world view, perspective,...). We/they will be more frustrated and out of sync if not. There is great room for conflict. Individually, this ultimately means finding the right compatibility for your optimal system based on your personal conditions and the flexibility to update on a regular basis.

The compatibility and updating is something that I work on a lot with clients, as the shift is consistent, but the required mindset and openness is the more critical work. The behavioral change is a natural expression of the inner shift.

One simple way to ease yourself into preparing for a system upgrade is taking a moment to recognize the technology that enhances your life and expressing gratitude with it on a regular basis. Before logging off, simply say, goodnight, and thank you. You are on your way to a better relationship already.

Chapter 4
Digital Self

Which digital self do you identify with most? Are you digital averse, resistant, cautious, balanced, curious, hoarder or addicted? Before I go into the theory and how the various profiles evolve through the stages of social optimization, I suggest you pick which one of these you are. I will give you a little description so you can help identify yourself. Just look at the summary image and see if you can determine where you identify yourself most. Where we ultimately want to be is that balance point. It is going to change with different things that occur. But if we help ourselves really get to a place of awareness or conscious evolution, we're going to get to digital balance faster. It is at this state that we are at peace with technology and become more fluid in our relationship. This is less about your evolution and it's more about your relationship with the technology and your relationship with those triggers.

The Profiles

The Profiles

If you recognize yourself as *Digitally Averse*, you might be underestimating your profile. I say this, particularly as the nature of this book being launched as an ebook, you are either consuming it on a e-reader or even a smartphone. Even if you are one of those wonderful humans that still loves the printed page and purchased a physical book, this is often just a preference. Nevertheless, if you even consider yourself as an online entrepreneur, by the nature of your interest or even recognizing that online is a possibility for you, expresses your profile as more likely to be resistant than averse, as this is the extreme. That said, it is critical to recognize that your audience may be Digitally Averse. In which case, by virtue of only providing digital delivery of your product or service excludes them. It's not just about you, because we're working with people. Without people there is no audience, which means no money coming in, because there is nobody to pay for your services. If your are delivering a product or service digitally, you also need to be very aware of how the people are interacting with technology that you're delivering through.

Even though this one most likely doesn't really apply to you, it's important to recognize with the Digitally Averse the sort of fear anxiety-based reaction. They proclaim themselves Luddites. Do any of you have clients proclaiming to be Luddites, yet they have a smart phone? That's a good way to qualify them right away. Let them know that they are much better off than they think they are. Alternatively, the true digital averse may not have WiFi. Their connection is provided through phone or cable connection, they feel they have no control of that connection. Often its because they're afraid the air, that space they do not know or understand and cannot see. The air has got all this technology, and they cannot control what's happening in it.

The *Digitally Resistant* feel a very strong dependence on techie people. They can't do it without someone sitting there walking through the process. They're often driven by privacy, safety, and health concerns sometimes even EMF concerns. EMFs are electromagnetic frequencies of which some people are very sensitive to, others it's more psychological. For some it is psychosomatic and they are simply afraid of the air. Although, most likely, none of you fit into that category awareness of its existence and how it can be an issue for your clients is important.

The Digitally Resistant insist upon a minimalist approach to technology and social media. They limit the reach of their audience, who need and want to hear from them. They limit their clients' access both in speed and content relevance. Information often misses being released just in time which improves content relevance. The result is feeling lost and behind. Like all the stages, it is possible to shift profile based on large shifts in the social optimization stages. The Digitally Resistant is particularly susceptible to the shift (either to Averse or Cautious) as the impact on their business can be quite extreme. From the Digitally Resistant we see those examples of people who "quit" social media when triggered. It is quite possible to have come from Digital Resistant to evolve to any of the stages based on positive experience and support that has enabled the resistant to appreciate the improved quality of life and work.

The *Digitally Cautious* try new tools with hesitation. They are open to observing the benefits and maybe try with distance. Willing to try technology that is proven and easy to adopt to. They are often categorized as a Late Adopters. They need some handholding assistance to navigate the opportunities to connect and grow business. There's a lot of excess time dedicated to learning and trying to understand rather than applying. There is still room to move.

I will come back to the *Digitally Balanced* as this is the middle ground we strive to reside in. Let's just say for now, that even the Digitally Balanced can be affected by triggers. This is important to remember as with the social optimization model, the digital era ensures us of one thing, constant change of conditions and environment. All other profiles have the potential to have moments in which they feel they have attained the characteristics of Digitally Balanced.

The *Digitally Curious* are similar to Digitally Cautious, but more exploratory. They collect tools and take deep dives into learning them. This can often result in a loss of focus on high value actions and relationships. They desire to maximize benefits of digital tools but are unclear on how to manage them. Basically, there's a lot of opportunity there. It requires a shift to recognizing where your behaviors are coming from and how it's impacting your business, to create room for movement towards balance. Often delegation of certain technology in order to scale comes into play here. Just because you can do it, doesn't mean it is the best use of your resources. After all, as an entrepreneur in particular, your time=money both wasted and gained.

The *Digital Hoarder* jumps at any new technology to solve the problem at hand, distracting them from high value actions and relationships. They are early adopters of anything new and shiny in the digital landscape. This behavior often results in a large portion of both time and money resources wasted on digital consumption. Even if you do not relate to this, think about the people that you're working with. They may be your clients, customers, partners, suppliers. Some of these characteristics may apply and hinder your potential working relationship. It's also something that you can help them to be mindful of. If you recognize this in yourself, you might have a buddy check in with you. Conscious technology consumption can create surprising efficiencies in many areas of work and life.

The *Digital Addict* is on the other end of the spectrum from the Digital Averse. They are the one's that drive the statistics on smart phone usage. They cannot be without their devices and are driven by the dopamine rush of a new like or share of their content. Often this addictive behavior is not limited to digital. In the more extreme cases, digital addiction can result in a psychological diagnosis of nomophobia (fear of being away from mobile phone). Their usage is extremely detrimental to their relationships with clients and friends as well as their personal well-being.

Awareness helps to recognize what those triggers are and when they occur. Often in this case the triggers are not related to tech in itself, but rather are related to an internal struggle. Help is sometimes required to move to be coming Digitally Balanced. It is good to be mindful of not only your own tendencies in this direction, but also your clients.

We are all striving for *Digital Balance* where the relationship with technology is fluid and enhances the wellbeing, relationships and business. The Digitally Balanced have clear boundaries and guidelines for how and when technology fits into their life and business. They are careful to avoid overload by delegating that which does not require their personal attention. This balance leaves room for new ideas, innovation, creativity, and of course building new and maintaining existing relationships. The Digitally Balanced are skilled at growing their business both online and offline by using the online interactions to enhance the offline connections and vice versa.

The Digital Self Mastery series is intended as the beginning of a dialog. In an effort to continue the stories and learn from others, there is a Digital Self Mastery group on Facebook for readers to share their stories and ask questions. I will do a monthly Facebook Live to share new findings and connect with all of you over *fika* (Swedish tradition of connecting over coffee, but if you are on another timezone, feel free to join us with your beverage of choice). The group is open, so feel free to join us and share it with others who you think would be interested in this conversation. I'm collecting stories for future editions of Digital Self Mastery. If you have a story relating to the digital self that you would like to share, please do not hesitate to message me in the Facebook group.

Sure, I expect that you're going to be moving around. Consider what is your natural behavior around technology. You may relate to two different sides of balance as you read the descriptions. This is partly because your relationship and the nature of the technology is evolving. Participants in my programs often begin further out to the side of the Digital Self spectrum and evolve closer to the balance point with smaller swings when triggers occur. It all depends on your level of comfort with the technologies. Increased comfort and peace affords you a greater willingness to try other things that may be bigger leaps. Your relationship may never be fully sustained in balance. You're going to be moving up and down. Ideally, you will arrive at a state where your swings in the pendulum and your spirals in the stages are smaller. The new technology becomes the old as it is adopted. This shift will result in more complex profiles. The movement is expected especially with big shifts in widely adopted technologies.

The iPhone launched in 2007, changing the way we thought about "phones." When it first came out, how many of you were the first, and how many resistant? Now, could you live without a smartphone? Our kids have come of age in this period and don't know life without them. We pay for transportation, order coffee, keep track of people and things and of course chat, text and share. We do all this with a small device that is more powerful than the computer that took men to the moon.

Within a ten year period our relationship with technology has been completely transformed. Across the globe, that transformation has even reached areas that 10 years ago did not even have Internet access. This is most apparent with the recent report of smartphone penetration even in Myanmar reaching 90% access. So much for that fancy new fax machine that made my job during college seem cutting edge but now sits utterly obsolete as I can scan and send a pdf directly from my smartphone. Wow, just thinking about that makes me overwhelmed with gratitude for the power of what I hold in my hand.

The shifts will be triggered as other such technologies and new innovations arrive. If we do not heal that relationship to allow it to become more fluid, we are less likely to reach the balance point at any given time. We will be more inclined to wider swings and deeper falls. We must evolve together with our operating systems or be left behind.

Evolving

Let's look deeper into this evolution and how it fits with the stages of social optimization. Keeping your digital self persona in mind as you go through this, I want you to think about how experiences might have triggered you at the different social optimization stages.

What might have been a trigger that caused you to shift in your stage or your relationship?

Triggers are the experiences that like kairos and metanoia, referred to in chapter two, either consciously or unconsciously result in a change of social optimization stage.

Let's take a common example, your computer crashes in the middle of something very important, losing all your data and all your hard work. Most people have experienced that at some time.

It would be a classic trigger but what we are looking at is not the computer crash, but rather what was your response?

Did you respond by replacing the computer because, "I don't trust this computer anymore?"

This of course, is a huge expense in both money and time invested in set up of new and recovery of as much as possible from the old, and often not necessary.

Or did you seek help?

What was your response there?

Did you panic to the point where you didn't want to touch technology?

Did you buy new backup systems?

Or did you have backup in place?

Do you remember the wellbeing consequences of the crash?

Did you physically feel the stress?

How did this display in your body (headache, depression, elevated heart rate)?

Were you able to concentrate and remain focused and present with patience?

How did this impact your work?

The first social optimization stage in which we question or rise in *Trust* is the stage of *Hope*. Think about your first experience acknowledging the presence of technology and the Internet of Things. The need was based on education, entertainment, task-based interaction.

Think about your first mobile phone, you might have used it for making and receiving phone calls maybe some text, SMS, or the more advanced media messages in small sized. The challenge is that the technology is great when it works… frustration at the stage can lead to a total dissociation later.

How you approach technology can be based on early experiences, often forgotten amidst the rapid emergence of the digital era.

Was your first real personal technology experience with a smartphone?

Or do you remember your first time using computer?

Perhaps you had the more intimate experience of a wearable fitness device that is tracking your health data.

Whatever the initial interaction, it began the evolution of your relationship with technology. It is from there that we can establish our baseline and identify what triggers can help us evolve to the highest level of social optimization with our technology as our ally.

We strive to reach the higher levels of evolution. At the *Care* stage we achieve greater connection with our community expanding our impact. This is motivated by *generativity* and the interest in leaving a legacy. Ultimately we hope to rise to *Wisdom* stage. The expansiveness of ego-integrity enables not only our work to have a greater impact but also to co-create with other changemakers for sustainable solutions that benefit the greater good. The closer we are to *Digitally Balanced*, the greater potential for sustaining these higher stages.

In this next section, I will share with you some stories from individuals in each of the digital self profiles. These stories are a combination of interviews, clients stories and experiences. They provide perspective on how our relationship with technology evolves, as can our digital self if we are willing to evolve with it.

Part Two:
The Who..

Chapter 5
Digital Averse

Sheryl Bernstein is a creative clarity coach, whom I met through Lisa Sasevich's Sassy mastermind. It was such a rich interview that it is largely unedited. I was so struck by her evolution that it seemed worth sharing it in its entirety. (Sheryl's story is in italics)

- *I am a Creative Clarity Coach. I coach in the Law of Attraction and also have a background of over three decades in showbiz. I combine the two. So I can help people be creative; come out of their shell, get over their fears, move forward* whatever creative endeavor they are hoping to do. I believe, I was the last person to go kicking and screaming into getting a computer.

- It was 1992. I was in my house and I was like, "I'm fine, you know, I like my little day book." I still have it, with all the phone numbers and everything, just in case. I still am thinking, oh, you know, that computers go down. So, I was the last person to get a computer. Then I was flabbergasted. What do you do? It took me a while to learn.

- I think I was also the last person to get an iPhone. Now I have a nice big, whatever the new one is. Love it. It took me while to get used to getting into the computer age. At first, all I did was email. Little by little, now I'm totally loving it. I'm anxious to get guidance on the mastery of all there is from Facebook. I know there still a ton of stuff and strategy to learn about. Just this morning I was thinking, "how do I actually give somebody my address?" I know there's strategy still to learn.

- Joining the Sassy mastermind was a big boost in getting really tech savvy. When you are presented with a huge website, full of tabs, each tab has hundreds of different elements to it, you learn how to navigate. You learn how to use it. At first it was very frightening. Now, whenever I am presented with something like that, I'm like, "OK, it's do-able."

- My coaching mostly is on the phone, but still it's through a phone that has a conference line. All they have to do is call in. But just giving somebody a call-in number and access code, if they are not familiar with what this is, it can throw them off. Some coaching clients are like, I'm not text savvy, I don't want to do any of that.

- Here's how it even colored what program I thought of developing. I wanted to I work with women who are 60+ who have retired. This is just one idea I had. They are retired and wondering what they want to do. Some of them have no idea how to use a computer. No idea how to do any of these things, email even. They are afraid of it. They don't want private messaging even. For people already on computer, they don't want to add another thing. It can color how I think about "What am I going to develop? Who do I want to work with? How do I want to do it?".

- My once tech anxiety, fear and resistance has turned into, "Now I'm an explorer. Now, I'm an adventurer. What else can I do?" I know there's like a zillion things I don't know about this thing, my iPhone. As far as my computer itself, I am getting into creating a lot of stuff using illustrative systems. There is still this feeling of, "I only know like 10% of what is possible." I am pretty good, but I know that there's so much more.

- The anxiety still comes up. Now, I'm just going to ask somebody how to do this. I'm going to find somebody to tell me. I have since joined a techie mastermind where I can get all my questions answered, like that! I post a question, and the answer is in there within a day. I can see that learning the technical part of the digital world will advance me so much further than I can even know. It will streamline. It will make things faster. I know I do things in some kind of a prehistoric way sometimes, cutting and pasting and going backwards. I know there's just a button to click somewhere.

- I have delegated creating my little eBook to a virtual assistant (VA). Everything else, up till now, I do to myself. I'm thinking if I get into a large launch, or something a greater nature than I am doing now, that involves a huge amount of emails going out, connecting to different platforms, I would get help for that. Right now, I do it on my own. I'm kind of a do-it-yourself gal.

- Perhaps not hiring someone to handle that is still a resistance of mine connected to a money mindset. The chicken and egg, which comes first? Do I up-level my business first financially and then get help? Or do I get help and that will make my business up-level financially? It's, of course, a question we all have.

- I know when it comes to something at my house or my car or fixing the air conditioner, I'm not going to do it myself. I'll call somebody who, hopefully I have gotten the best pro, to do it. Part of me is like, "Why don't you do that in your business?" Somebody who, that's what they love, love to do. It takes them half the time, sometimes in my case it could be like the eighth of the time. I think that the further we go in our business, the more we think that *it becomes a possibility to entertain rather than, "oh, no that's years away."*

(Heidi) Have you ever had a traumatic experience with technology that set you back? It sounds like you've had really positive development overtime. Is there anything, could be traumatic or something that was monumental. You said that once you got over your aversion or resistance, and started working with the computer, all sudden you discovered all these efficiencies. It just opened up a the world and built your confidence that much more. You're clearly open to trying other things. Nonetheless, occasionally there's something in your evolution in your relationship with technology that may set you back where you're lose trust. Do you recall anything that might have caused that kind of a trigger?

- *I'm sure it's the same story many have had, where I worked on documents for a long time and then I lost it. I don't even know how I lost it. The computer shut down and it might still be somewhere.*

- *I also do voiceovers and I now do this from home. In the last decade or so, it went from going out to studios, where the engineer's in the studio and all I need to do is talk into a microphone. Now it's all home-based. Everybody does it at home now. I have my microphone and a system. It doesn't happen very often now, because I know to save. "Save" is my friend. I will maybe have worked on something for quite some time, recording and even editing and something happens. The computer might crash. I didn't have it saved. It's gone. Those are the kinds of stories of where working on something and my naiveté of how to protect myself wasn't good. I didn't know how to do that. So I've lost something I've spent a lot of time on.*

- *There's also email going to the wrong person. Auto address, sometimes I'll start writing you, Heidi, and I have another Heidi-friend who's address comes up, but I don't catch it before it is sent. Perhaps there is tension between the two or they know each other. Maybe I send something meant for just for you. In any case, I never say anything really bad, so it is not such a big problem. Its that auto... when the computer starts doing things ahead of me, it anticipates what I am wanting and it's anticipating incorrectly. That will maybe cause a problem. I haven't had a huge, terrible disaster. My path has been more resistance, anxiety, fear transformed into joy. Let's see what this can do. Let's take this baby out driving.*

(Heidi) I would say you are now Digitally Curious, gliding back and forth to Balance. You're at your start of this and there's only more to come. Curious resides ahead of Digitally Balanced. I would say you evolved quickly past Digitally Cautious, where you limited your access to new clients and growing your business. Digitally Curious is on either side of Balance. One of them says, "I need the stop now, even though I should be trying it but its too scary" and the other is past it to the point of, "this is really exciting I want to try it all, but maybe there's not enough time to learn all."

- *I learned what I need as I go. Do something and then see what you need. There is so much, so very much. My anxiety in the beginning was so forthright. "I don't need that!" It was beyond resistance. It was indignation.*

(Heidi) You have evolved beautifully from Digitally Averse or Resistant to Digitally Curious. It's a big move to go all the way to Digitally Curious from where you started. Congratulations! Because that's really going to help your business. Its pretty exciting and not easy, so, well done.

- *That's the piece I think we need in every department. As an entrepreneur or anybody, to be curious and go jump in and then see, "OK, what do I need to do this? Who do I need to help me? What do I need to know?" and move. Otherwise, we can sit here for just trying to learn everything. It advances so much faster than we can. I know there's more happening as we speak.*

(Heidi) Have you ever had any experiences were you felt that the technology enabled you to be more generative? Where it expanded your reach in your business and connected you to people who are doing similar or complementary work because of your access or your curiosity about technology?

- *I can think of things I do every day or every other day, just Facebook LIVEs that have come out. I love doing that. That's my background. I love to be on camera, to be able to speak. So Facebook LIVE is a whole new world. Every platform now is coming out with something. Webinar jam is going to have some kind of live thing. Zoom, like we are using is amazing. So the connection of Facebook LIVE and putting video out there, in that moment, as I push that* button… There is something so magical about that. That connection where you being very real, you're being right there. So, that. Just having an email platforms where I can generate emails to my list. I mean, how great is that?! I still create it every time. I put it out. I am now in the process of learning how to automate a sequence. That's my next stop. I know that seems like a lot.

- Some people say they don't get what they need right now, they get what they are going to be growing into. Email sequencing, Facebook live, I mean I'm sure there's a lot of things that happened all the time. Just emailing people, private messaging, private messaging a whole bunch at once, the one to many abilities are just great for me. I know there's I don't even I know that I could be doing. I see the big guys doing webinars where they're standing next to the screen.. the screen is showing something… they're talking in a little square in the corner… I want to learn how to do all that.

(Heidi) That's fun stuff. Sometimes it's good to get somebody else do it for you. Then you just share your content in their format.

- That is something too, finding that person. There is the anxiety again. Finding the right people is a hole that can be a job in itself.

- The thing is, I'm all about that nature and flow, the beauty of the outside world, flowers and plants. You want to get the people connection. I like to get a nature connection *into the technology. That's my THING. You can sort of do it. Sometimes the illustrations don't quite get what I want. But, technology is here. This is where we are.*

Chapter 6
Digital Resistant

Standing on the stage speaking about technology looking at a sea of faces with expressions ranging from excitement to fear and panic is what lead me to developing programs to improve our relationship with technology. The strained postures, the looks of defeat, the raised eyebrows and deep sighs are all visible response in the faces of the Digitally Resistant. They have a powerful message or service to provide. They are limiting its reach by not managing their relationship with the technology that can provide that reach. They know it, too. They may not perceive it as a relationship issue, but their faces tell another story.

I want you to think of your clients that might be stuck in this profile. As this is the Online Entrepreneurs edition of Digital Self Mastery, I assume you are delivering your product or service online. Therefore your clients, whether they have a good relationship with technology or not, need to access your products or services through whatever online tools you use. Their resistance may cause an unnecessary relationship to your program. It is important to recognize where their reaction comes from. Is it the content or the means of delivery that they are reacting to? Be mindful of this relationship when doing discovery calls with clients to be sure you/your tech are the right match for them. Otherwise they may not be able to get the outcome that you are promising.

After interviewing a tech coach, she sought me out to share a client anecdote. She delivers her program over Facebook Live, and uses a Facebook group for community and communications. Her client reached out to her and said she hadn't received any of the content. It turns out this person had never logged into the Facebook group and was resistant to using it at all. This was clearly not a good match.

Recently at a meeting of 300+ online entrepreneurs, I watched those same expressions appear as a technology company presented the "back-end" to deliver and manage online programs and lists. There was, of course, a degree of the response related to the investment. As Sheryl shared the challenge of knowing when do you go from DIY to team and software to manage it all. Growth always has its challenges.

The Resistant struggle with this step, which feels more like a leap of faith. (The Hoarder, on the other hand, may take this step before they are ready and spend excess time trying to manage it themselves, resulting in stagnation from delayed release of content).

If you are providing technology services (that person they delegate to and trust), make sure you find a way communicate in their language both when listening and when delivering solutions. Miscommunication between the Digital Hoarder and the Digital Resistant is very common. With a little effort and understanding this can be a very successful partnership, win-win on both sides. Building trust and well documented communication can avoid a trigger to tumble down the stages.

The Digitally Resistant often fear the unknown, whether that is loss of control, time, space, or even (as mentioned before) the air. That resistance keeps them using old systems that serve them, but not efficiently. Sure, Outlook or basic mail systems work. How much extra effort is put into tracking unopened or bounced mails, groups, different program or client groups, active and former clients, etc…? Are they even mixed with personal mail? Just email alone can require a full-time assistant.

Before taking the leap from individual face-to-face clients, the Digitally Resistant face the challenge of mindset. Their fears often share the theme of loss; their personal touch, control of their content, their brand and even competition. And yet, when you extract the elements that are foundational, do not require personal touch, provide insight into deeper work and still give value, you have the pieces to deliver online content to the masses. It increases the value of your personal touch, 1:1 or small group work. The technology is simply the mediated means for delivering your content.

Either learning the basic tools for delivery or finding a good person to delegate it to, becomes an easier task with self-awareness. This is where communication is key, as mentioned before. Honesty with your team as to what you are capable of, have bandwidth for, want to achieve, and are willing to do is critical. If you are a "team" of one, as many online entrepreneurs are when they start out, you still need to be honest with your team. Do yourself a favor and find a mentor. Finding the support to make the rest happen will be easier.

Chapter 7
Digital Cautious

Perhaps the most fascinating part of writing this book has been interviewing people close to me. My mum, Tally Forbes, is an artist as well as a retired executive from the non-profit world. I have watched her relationship with technology evolve over the years. Her curiosity never ceases to amaze me. She approaches it with caution, but always a glimmer of curiosity. This comes across in her approach to most things from art to business. Her most recent adventure has been in sharing her art outside of physical gallery spaces.

- I post my paintings on Facebook and ask for comments. This gives me support from Facebook friends in my mostly solo world in the studio. I have sold a few this way too. It feels like I'm in a community of like-minded appreciators of art.

I was lucky enough to have her join in one of my early *Digital Life Balance* programs. As an adult child that has lived mostly abroad on the opposite coast from my mother, it was a great experience to see through her eyes.

- I am retired and focusing my life in two areas – painting and participating in monitoring international philanthropy project effectiveness as part of a community of philanthropists in Boston. The balance of the introspective process of moving the paint brush along a canvas to express myself, and connecting to the world through reading, researching and attending discussions in person and online creates a perfect life balance for me. In both areas, I use the internet and various applications to share and learn on a daily basis.

- Although I occasionally sell paintings as a result of posting them on Facebook, I use technology primarily to participate in community discussion on the painting process and to maintain friendships.

- Many of my friends and I enjoy keeping track of our grandkids on Facebook and text. As we find grandkids are "of few words" if we send a text we can at least get a short response out of them. Many times the response is *simply a smiling face or some other silly image that I have never seen before and could not find myself, but I do feel connected to them through this process."*

- *I love the photos that they post on Facebook when they are being silly with their friends, when they have won a game or race, or when they have been in a concert or theatrical production. They are too busy to write a letter, but never to busy to post a photo on Facebook. I believe that my relationships with grandkids is more active by using these technologies.*

I can contest to her active engagement across generations and communities, including with her art. Her inquiry posts with her latest creations receive insightful and thoughtful critique from grandkids, extended family, friends, and art lovers across the globe. It is exciting to see the interest and engagement in response to open and creative inquiry.

One of the favorite comments that ever came out of my mother's mouth was, "We are too polite to our devices. They don't understand us, unless we are rude." She kept telling her friends to stop being polite. Chuckling a bit to myself, of course, I had to ask her to explain. But it made total sense. My mother uses Siri and voice commands for Google a lot. Using the words, "Would you mind terribly looking up the directions to Trader Joe's, please? As opposed to, "directions to Trader Joes." There is less room for misinterpretation. And quite frankly, Siri doesn't care if you ask politely.

In a way, I think she and her friends get a quite a kick out of being allowed to be rude, after a lifetime of New England restraint. Far be it from me for assuming that. But perhaps it is their way to throw caution to the wind.

Often there is a swing between Digitally Cautious and Digitally Curious depending on the experiences and the support received.

Michelle's, Visibility Mindset Mentor, story shows a bit of the pendulum as it can occur.

- I have a huge story that getting up to speed with new technology is going to be hard and take a ton of time I don't have. I often think, "There's probably an app for that…" but taking the time to research it and then trying to decide if it's worth the time to get up to speed using it is often a "no" for me.

- I've learned that it's easier for me to learn new technology with someone who already knows it and have even started being more patient about getting to the bottom of tech challenges via the articles and help forums available.

- For example, I use a calendaring app to schedule client appointments. It syncs with my google calendar so I can have everything in one place. The sync stopped working. So all of a sudden I had clients double-booking. This happened a few times before I realized what was going on. It felt totally overwhelming!

- I couldn't send out scheduling links without triple checking them and then troubleshooting why time was showing *available when I wasn't. Syncing and unsyncing – my patience was sinking!*

- I resented the time it was taking to check and manually schedule people…I resented the time it would take to weed through all the help articles to try to figure it out…I resented the fact I couldn't get a person to help me! But I realized this was my own impatience…I figured it was easier to work it out with a system I already know (and have loved up until this point) than to get up to speed on a whole new system.

- I emailed the tech support.

- It was partly their problem…and partly google's… and partly mine ;)

- They fixed their end, helped me understand what was going on with my Google calendar and what I could do to prevent the problem…and gave me 6 months of free service!

- It was easy for me to be "frustrated with the technology!" Really, sometimes it feels like I can't live with it and I can't live without it! But realizing that what was causing me the most suffering was my own frustration, impatience and resentment was very powerful, freeing and ultimately a total game-changer!

- Sometimes the tech is easy…I download an app and it works! And I love it. I'm always surprised when that *happens! I love my mindfulness app...it chimes. I breathe. It was easy and fluid to figure out...so the user experience is a huge influencer for me!*

Chapter 8
Digital Balanced

The year before I moved to the US from Sweden, I made a new friend, Malin. We were both deep in the mire of balancing managing our businesses, family, health and finishing our PhDs. Although at the time, I interviewed her for my book *BE-ing@Work*, I found the content of our interview applied well for *Digital Self Mastery*.

> **fika** [²fiːka] taking time out for connection and coffee (Swedish tradition)

Malin and I often met for *fika* and surviving the doctoral process talk. Fika is one thing I really miss from Sweden, and perhaps why I do a virtual version in my programs. Not only was the coffee great, but the opportunity to put down devices, large and small, sit across from one another and connect was inspiring. So many great ideas stemmed from these conversations. But also the personal connection allowed us to have a connection that translated when continued in the virtual context.

Her consulting company, MOOV, works with design thinking, inspiration, ideation and the implementation phases. She relies on technology for project management, collaboration and team management, client meetings, as well as (at that time) for her doctoral research, from OneNote to dropbox and Skype. She conducted interviews and feedback sessions via Skype.

I found it particularly striking that she emphasized the power of immediate positive feedback using text, saying, "THANK YOU" or "HELLO" after an exchange. She insisted that it was critical not to keep a "scorecard" of invitations and "thanks." Instead to focus on giving rather than receiving or the balance of reciprocity. The implications being that without a method for immediate response, the instant message, the value of the message and relevance of it was lost. We have become increasingly reliant on technology to provide response without delay, our attention span is diminished, even for gratitude.

A Digitally Balanced Self, can fluctuate with shifts in either direction. That said, for the few that I have come across that start in a balanced relationship with technology, it appears to shift less. Usually the trigger is not related to technology but instead an external or internal experience. Personal health often is a clear and unexpected trigger to re-align.

When your brain starts sending you clear messages to manage your neurotransmitters that keep you focused, happy and healthy, you listen. Cortisol spikes from stress that lead your body to adrenal fatigue. Even device notifications can be a stress trigger. Blue screens disrupt the melatonin and dopamine which in turn disrupts both falling asleep and staying asleep. Devices in bed can minimize intimacy with partners, limiting both touch and the needed oxytocin touch provides your system. Basically, out of desperate need for self care, we create boundaries and rules for devices in the bedroom.

You may have noticed that I refer to "we". Like Malin, it was through personal wellbeing challenges in the last several years that I had to learn different ways to relate to my technology. This does not mean that technology had a negative impact, actually quite the contrary. Auto-immune disorders triggered all kinds of issues for me that impacted everything from my heart, to my head and my extremities. Looking to technology to help compensate and track or manage the newly limited energy and resources that I had, I found a friend and ally. I do not think this would have been possible without a balanced relationship with it.

I like the way Malin emphasized, "writing your dissertation on 15 minutes a day" using a timer. Sometimes we have to work within our limitations and sometimes we can stretch them. Fortunately, even when they are limited, I am inclined to believe, that technology stretches them a little bit more even when we can't. I can't be all that off, considering we both eventually did finish our PhDs and are back to work we love. (We are still standing, I might add.)

Perhaps we are able to remain balanced because of our Swedish schedules, taking 10 weeks off a year. That does not mean we are offline 10 weeks, but rather we have built our businesses to full throttle the rest of the year and limited demands during holidays and Summer. After all, as long as there is an Internet connection, we can still work enough. Or rather, should I say *lagom* work.

> **lagom** [ˈlɑ̀ːɡɔm] Just the right amount *(Swedish)*

Since moving to California, I established two regular walk-and-talk friends that I try to connect with at least twice a year for what might be considered the Silicon Valley version of fika. One works as a Director at Apple University and the other was, until recently, the Chief Game Designer at Google (he's now back to developing serious games for social impact). I loved these walk and talks for the conversation, inspiration and humanity of them. They both work at high levels in immersive tech environments. And yet, they exhibit a healthy balance of respect, appreciation and curiosity about tech now and in the future. I always walk away feeling confident there were still good people in the tech ecosystem influencing the developments of the future.

Chapter 9
Digital Curious

Sharon Svenson of Svenson Hypnosis shared her story with me over Zoom earlier this year. I love the way she fits into Digitally Curious with a nice evolution towards Balance.

- *I started off in 2006. I worked with the typical reasons that people come to a hypnotist which is stop smoking, weight loss, that kind of thing. Right from the beginning, I got some very challenging cases where people came for pain control. I kind of branch out into helping people with incurable illnesses. I got to see what was possible and to see illnesses turn around.*

- *Fast forward to my business today, now I'm gearing up to work more with small entrepreneurs, coaches, healers, that kind of thing, helping them to connect up and jump past their limited mindset into their big dreams. (I help them) to overcome those fears and anxieties and technology is one of them.*

What particular story did you have into mind when I asked for interviews? Your own experience with maybe going online with your business? The third piece is how you feel using a mediated tool for communicating with your clients, how that affects your relationship or ability to deliver your content and your clients response?

- *I wanted to get bigger and it felt like, in order to do that I had to get comfortable with technology. Up to that point, I would print flyers around my town and by word of mouth my business grew. If I wanted to go bigger, like doing group programs, that kind of thing, I knew that I had to change things. If I wanted to deliver more to the clients too that I really needed to go online.*

- *I had very much a fear of technology. I think a big part of it was, I was afraid what it would do to me. I didn't wanted to take me away from my joy for life. That's a big part my programs. I call my programs, the Mastery of Joy: Follow Your Joy and the Rest Follows. So, I was fearful* of that and also fearful everything would go blank. For instance, I had this fellow that was building my website. He tried to explain it to me and my mind just shut off, literately. OK, so, did I hear any of that? So, I was afraid that I wouldn't be able to really use the technology and make it work for me.

- I guess I tried to, "heal thyself," as they say, to be more open and receptive to this working for me. It led me to the CoachesConsole Boot Camp and they train you on their system which is brings all the different elements of your business together. So it has your opt in page, your website, the back-end so you can do your agreements. They do it piece by piece. I was like, "oh my gosh, piece by piece, I could get this. I could do this." It just opened me up. To me, I can't even tell you, it was like a miracle. It changed my business.

- I went from this ridiculous low levels and now I was able to offer them extra materials, online. It changed the success for the clients, too. Now they could have more in between support. I have lots of resources and backup material. When I do a sample session, I can instantly send an agreement to sign. Now working with people in different countries and different states became so much more easy. Where I came from, figuring things out on the Internet, it just felt so impossible.

- I love that I know now that it can be a beautiful gift when used in a very selective and thoughtful way. I loved the articles that you posted that I just read this morning.

You don't let it take over your life. Like people that have the phone that are just constantly attached to them, or Facebook. I'm very conscious of, "OK, I'm only going to spend so much time, and I'm going to be selective of what groups I participate in, and who I connect with too." To me it can be this beautiful gift. But like anything, just like alcohol, that you can use it from empowering place or from a place of lack. I'm in a much healthier state.

- Sometimes, I still going to that place of feeling angry with technology, that I have to continue to learn continue to figure things out. "What, they change things again?!"

- But, overall I feel like it's been a blessing. It keeps my mind active growing and evolving. It does seem sometimes like the technology part can maybe be a bigger part of my business than I want. My hope is to be able to I hire somebody that loves that stuff and get them to do. I don't have a really big list, but I feel like the people that I do have on my list want to be on it. I have good open rates and engagement. That feels really good. I have 300+. Sometimes it does not feel manageable, and sometimes it does.

- It's tough when I doing a new program. I'm bringing together this new group program for entrepreneurs and coaches. I still have year long programs, pretty much. I want to make sure that I'm really there for the clients. With the programs I'm creating and started to do partnerships with other people, sometimes it can feel overwhelming. Trying to keep up with all the technology pieces.

- *The big part of my packages, the transformation is through guided meditation. The coaching I do for the CoachesConsole system, it's just a side thing. It's nice because I get a lot of connections and meet a lot of interesting people.*

- *(Client response to immediate delivery) I have one woman that I'm working with. She comes in person. She's local. I first tried to train her, "bring your iPad and we'll work together." We did like 5 or 6 tutorials. Ultimately, it was not working for her. So I ended up burning the recordings for her, because I record all that sessions. I give her everything in hard copy. I print it off for her things, the things I want to have. There's a few people. I want it to work for them. There's a few people that it just doesn't work for them. Its not at that point. For the most part, I found that people are pretty thrilled with that extra addition and support that they can have. This other woman, she likes the recordings that I send her. But, she doesn't utilize all the resources that she can. Yet the ones that do, it's available for them, which is great.*

- *I think of people, like anything, use it to whatever advantage they want. It's only a small portion overall. I've got one woman, she does Facebook but she won't interact with the email or other pieces. So, sometimes it can I be more challenging. In the old days you just call them up on the phone. "Here's a reminder, come to your session." Now, it's like some people they do the text, or they do Facebook or they just do email.*

- I think I have to re-adjust. Like always, get up to speed with the new ways of how want to connect. So, right now it's automated so that I have email notifications go out The ones that use Facebook or need to be notified by text, I don't have that automated yet. Its been a blessing to have the automation of the emails. Technology has really help me. It allowed me not to keep track of so much stuff too.

- I love having the reminders for myself. "Remember you have a meeting tomorrow," which I appreciate, but not everybody does that. So, usually I put it in my system to remind myself too. There's so many pieces that to have technology help you keep track of it is great. It feels like it's way more professional if I get a reminder. I think that really helped for my business. Even though I was getting amazing results, it took me to a new level of professionalism because people saw me in a different way. It gave me more of a legitimacy, too.

(Your relationship with technology on a personal level) - When I did that first year with a CoachesConsole, building my website etc, the whole thing got out of control. A lot of things like taking walks, doing yoga and just doing things that I really love went on the wayside. This has been my year of just being healthy and balanced and still moving toward on the technology. I remind myself that it's OK. That it's going to unfold incrementally. I don't have to get everything done at once.

- I don't really feel like I'm overwhelmed. I have different guidelines, like with Facebook participation, if I don't really connect, or if it doesn't speak with my heart, so to speak and resonate with me, I don't engage. That feels really good. I don't feel like I have to just engaged or just "like this." I feel free to do it on my terms. I'm not going to like it if I don't really know what it's about. That feels good.

- I don't have, for instance, my smart phone on me. I selectively answer it when I want to. I don't have it dictate my life. I have it set so it's not like, "ding, ding, ding." It's on silent. Unless I have got a client that is going to call. Then I turn the ringer on. At night, after a certain time, I don't do different technology things, whether it's just social media or working on the website. I try to carve out the spaces in my life so that I take walks. I do my yoga and keep that in balance. I have time for friendships and talking to people. That's important to me.

(Self care technology)

- I have this app called Gaia. It's brought a ton of good things. I do the yoga. We live in a really small town, so I feel like I can get these awesome teachers online and learn from them. So, I really appreciate and enjoy got that.

- I like the CoachesConsole system I went through. That's how we did the learning, through the webinars and engagement online. I think I wanted to continue to be with the CoachesConsole there they're having us do more as coaches. I want to be more selective in the meetings that I do and the people that I engage with. Connecting with others, I think it's a real blessing. I've met some friends and connected with them online. We supported each other as coaches, and that's been really nice. I did different mastermind groups with it. I feel like it's been a nice blessing in my life, too. It's just one that I need to keep in balance. It's way too many emails. That drives me crazy.

- Off and on I will go through and unsubscribe to different things. I'll go in at least twice a day and just delete stuff I don't want to look at. Then I've got categories I can put them into, when I don't really know if I want to look at it or not. Chances are, I probably won't. At least I have an option and I can get it out of my inbox. I can just focus on the things that I need to prioritize and focus on. Sometimes it just seems overwhelming with the emails. It's too many things to subscribe and focus on. I think people want you just to subscribe but they don't give you anything. They're just promoting other people. It kind of drives me crazy. Especially where there's no real engagement. I think it's going a different way maybe that email is going to be in the past. It (the future) is going to be more selective like with Twitter or Facebook groups.

Chapter 10
Digital Hoarder

I met Abby Kanarek when I was speaking at the Consumer Electronics Show (CES) in 2015 on Women in Wearables. Abby Kanarek is the operations or organization person for Living in Digital Times, they run much of the behind the scenes for some of the largest tech conferences in the world, CES for one. The nature of her work involves much of what mine did, at the time, testing and trying all the latest digital health gadgets and tech in order to stay up to date.

As she spoke about being an early adopter with a tech graveyard of sorts, a drawer full of technology that is no longer in use, outlived its value. The only relief was completely unplugging.

For her, and admitted stress junky, she has had to learn from the signals from her body. For something like conference management, there are many things that can go wrong, and most of those are related to people, not tech.

She felt she was at her best when the show was under way, not the lead up to. Also best when confident in the team. This is a challenge with a lot of part time people due to the nature of conference management. There is a lot of turn over. Just how much can be delegated and who to delegate to demands trust and patience. She chooses to "focus on end result instead of path to getting there." Each person has a different way of doing it. They begin with a team meeting at start, pre-apologise for snap commands, as they must keep flow. These kids are working for $20/hr, so she has to try to control flow of information for critical actions. Her role is to know everything that is going on: prepare and execute.

Women are looking to technology to help them improve efficiency. We also need to learn how to listen to the signals from our bodies. In Abby's case the signals can be drowned out by the chaos of the moment during the events. She anticipates complete burn-out and plans in post-event recovery.

Using activity monitoring devices at the event is gratifying as one covers a great deal of ground by foot. The reward features provide a sense of accomplishment without effort. Nonetheless, the devices end up in the tech graveyard post-event. They no longer serve a purpose.

You may wonder why I use this example in the case of the Digital Hoarder. I see her as an evolved Digital Hoarder, as she has learned to manage her energy and capacity by taking breaks. In her case it is more extreme as the events are short and intense but have long lead time and preparation. Many of you will choose to do live events as well as your online programs or services. In this case, being mindful of the toll it can take and the re-boot required upon completion can keep this instance from becoming a trigger.

Knowing how to read your body, or using tech that track your vital signs, can also ensure a successful event. Abby used the Muse to train her brain to meditate. What tools can you use? A meditation app is a good start. LiveAMoment, Headspace or the Muse device are all easily attainable. It is not over until the last curtain. If you are on stage, your brand is what you represent on that stage. Brilliance, not burnout, is how you should close the stage.

Recently, I received a re-connect note from her organization. I was happy to see it for two reasons. The first was the re-establishing the connection with like-minded peers doing exciting things. The second was, they were finally upgrading their list management technology. I felt a relief for their team. I thought of Abby and smiled. Maybe their team has grown, but even more so, now they are taking advantage of the systems that can support their growth.

Like the Digitally Resistant, the Digital Hoarder often stagnates as a result of limited bandwidth. They have "the lastest" most interesting thing "out-there." BUT they often do not have the bandwidth to fully take advantage of it, so other areas get compromised.

Here is where I admit my weakness, I swing to Digital Hoarder. I suppose it is partly a lingering effect of my geek girl self. I can do. But, that doesn't mean, I should do it. Of course, I can create posts using Photoshop to edit pictures, to create great backdrops that can have inspiring quotes overlaid on them (in a few days). Or, I can have my teenager whip up a dozen of them on Canva in an hour. Hmm, which do you think is more efficient?

Building a team, trusting them, and the technology are the hardest actions for a Digital Hoarder. Building in practices to rinse out ineffective processes or tools applies to any spring cleaning process. It is with this that I am able to get closer to Balance.

Chapter 11
Digital Addict

I met Troy Scott, author of Small Business Owners Guide to Digital Marketing, at a book launch workshop. Similar to my reaction to digital disconnect in this community, he said, "I found these folks in small business need structure and they don't have it. (My book) has that in five simple steps." To preface Troy's interview being placed in under Digital Addict, he is an evolved Digital Addict, as you will see in his story.

So, I asked him, "I'm sure you could give me all kinds of content that would be relevant regarding your client but what I'm really curious about is how you have experienced your relationship technology; where it has enhanced your life and your business, or experiences where you were hindered in the process."

- Upfront, technology has always been an enhancer for me because I am an early adopter. I'm a geek. So new stuff, new technology, I'm always going to play with it. That's the personal side of it. There's more to it as well.

- On the business side, technology is been a hindrance. I'm launching another company, technology had to be adopted in order for it to work. That slowed the process way down even though, as the creators, we (were out there saying), "here's how we can help you..." The technology was the hindrance.

- So, on a personal note, technology, I have embraced it. It's really helped me whether it's with scheduling things or managing projects or pure entertainment as another way to listen to music, communicate with podcasts.

- It's also gotten to the point to where I recognize that I do have to unplug. It's a conscious reason that I need to unplug. But also, technology is so part of my life now that if I don't consciously make it a point of "ok, it's Sunday, you're unplugging, you're not gonna work, not going to do anything, maybe golf, go out with the wife, go to *see a movie, do something that doesn't have to do with technology." That is a new habit, by the way.*

What triggered that habit? Was that you or was that from feedback you were getting from your wife or from people around you?

- *The latter, you know for sure my wife. But more over three cycles of health issues. The first cycle, no balance put me in the hospital. The true entrepreneur story. The second was trying to do too much and getting sick over and over. This third one was another case of just recognizing balance is super important. Technology was contributing to my being out of balance.*

(Heidi) What are your best tricks that help you re-find that balance but using technology? Have you used anything, at this point, or you have to shut it off entirely to get that balance? Are there tools that you use to create balance?

- *Yeah, I don't exactly remember the app, but I think it's Stride, the running app. So, having a structured walk is super important. No music. No nothing. Just go walk. Think. That's the structured part of my day. I strive to follow a loop and maintain that. For me, it is just completely dedicated to shut off. It's starting to become Sunday.*

(Heidi) How often do you do a digital elimination or spring cleaning? As an early adopter, we have a lot of devices and things. We try. We test. A lot of people are really bad about getting rid of the things that aren't working. So you can accumulate a lot of those. They're not really helpful to you.

You could maybe donate them to somebody else who needs them. Or you could maybe just move them out of the way. Is that ever an issue for you? It could be a physical product or a legacy software that, with investment in time and money, your systems would be more efficient.

- *Yes, It's coming in a couple different ways. Whether it is just physically getting rid of old, replacing components in the house, or looking in your garage and realizing you're piling up a bunch of unused technology. We had a recent experience moving house and into other house. My wife and I both realized how much junk we had that was technology related. Moving is great. Based on the way we plan to move into this new house, we left everything. We moved into this house with a couple of beds, a dining room table, and a couch (that we were supposed to give away). Because we're planning on cleaning the clutter. Oh man, it is so liberating. Unbelievably liberating.*

(Heidi) Do you do the same thing with your software?

- *I have. It has happened in two different ways for me. One, hard drive failure. The other is being is being cognizant of the pollution of bringing more apps into my world.*

(Heidi) For you to have this occur, it seems to require a catalyst: a move or a crash? Do you have systems built in to annually purge or survey of what is helpful or in use or not?

- *It is not innate. It would be super helpful to have something like that set up that once a year catalogs what is really needed here, especially on your phone. The Apps just grow. Forty percent of the Apps I don't use.*

(Heidi) Do you ever feel that things get lost in translation when working with clients that are not tech savvy?

- *Yes. That is a really good point. So, over the years I have realized, it actually happened at last night's dinner, what is perceived as casual communication in my world, is not translating very well at all. So I have been really making sure that if I am communicating anything that has to do with technology, I don't talk about technology. I talk about what can happen, the outcome.*

(On the client side) - *there's the case of, "I want it to do this." Their not having any understanding of that not being realistic. It is actually our fault. We need to be doing a better job at the process of asking the process. Documentation is great, but need communication that bridges the gap between intended outcome and possibility. That is usually the case. That is usually where it is broken. So it is not like, "what do you want your auto-responder to do?" The question is, "What are we doing with your customers" Or "what are we doing with your process before they are customers?"*

- *Regarding delegation, I came to the realization in 2015 that "I can't do it all." The innate feeling that you have as an entrepreneur, I think you have to get through that phase, you finally realize that it doesn't have to be perfect all the time. You can actually have somebody else, who may* even be better at it than you, do it. I realized that I could absolutely release the shackles of the burden of having to do it all. Which, interestingly enough, got back to what I talked about earlier about being out of balance trying to do it all. Huge benefits can be found.

- I came to the realization that I can't do it all. Its normal to think I shouldn't do it all, but I have to do it all cause nobody else is gonna do it. But I realized, I can't, because it doesn't make any sense. The net benefit of that was big increases in productivity and revenue. In a company that does over $10M a year, one of the efficiencies we outsourced realized an additional annual growth of $2.1M. That is a good example of getting out of your own way.

(Freeing up Sundays) - My wife is an entrepreneur as well. She is finding her balance again. I am trying to get structure back in. Living in San Jose in Silicon Valley, with my network the communication is usually around technology. I have another group of old friends that ride dirt bikes. I am the weirdo technology guy in that group. When we go out riding dirt bikes, that is another way for me to unplug. (As he said this, he lit up. Good on you Troy!) You know you have a sickness when you are like, "What is your click-through rate on that?" I've gotten much better at that.

Part Three:
The Where to..

Chapter 12
Evolving

Finding a mentor is a very effective way to evolve your digital self. When entering the world of online entrepreneurship as a practitioner, I had to find my people too. My first shift came through a push from my accountability mentor Simone Janssen. She then connected me with my online program mentor, Alina Vincent. Through my years of corporate training and teaching, I am keenly aware of the power of visuals. In the last few years, video became an increasingly easy and effective way to create and deliver content. I had to overcome my resistance to live video streaming. It was through these amazing women that I found my mentor that gave me a powerful presence for stage and video, Michelle Kopper. Finding the right mentor for you can make the difference between stagnation and elevation in your business.

Finding the support you need will level you up (or down, depending on the direction you need) to be Digitally Balanced. Whether it is a VA (shout out to Nicole who I could never do this without), a web design person, a social media guru, or even your tech savvy teen, having the support you need so that you can focus on your content, your gift, your message is critical. Being mindful of where your clients are in their relationship with technology will help you both reach a better balance point.

CLARITY

As we move towards a state where our relationship with technology has less friction, we can come to a place of gratitude, respect and curiosity. We remove it.

I challenge you to find something everyday that you can be grateful for because of technology. ...

> "Because of technology I am able to chat face to face with my father in a nursing home across the country, keeping us connected as a family despite the distance."

I challenge you to find something everyday that you can respect technology for that improves your life or your business...

> "The advances in battery and in processing power have enabled me to have a laptop and complete system that no longer hurts my back in my extensive business travel. Not only that, but with the cloud, I can access everything I have ever worked on when I am traveling, even if it isn't on my laptop."

The last challenge I leave you with today, is to be curious about the potential for where you can take your business and your desired lifestyle in the future, as a result of a peaceful relationship with technology.

What's Next?!

I do hope that it doesn't take our bodies sending us extreme signals of shut down to finally listen to the cues. Developing our own systems and finding the technology needed to support them is needed to achieve mastery of our digital selves. Each of you is unique in your story, your desires, your needs, your goals... Finding that which complements your unique you and your operating system is necessary of you to move closer to balance. It is possible, if you give it a chance.

We may never achieve a fully harmonious state of sociomateriality with our technology. But we can certainly strive for minimizing the friction and being accountable for where the friction derives from.

This book is used as a foundational element to the Digital Self Mastery™ Movement. It is intended as a starting point for the conversation. It was created as a tool to support my programs, coaching and consulting, so that we can to do deeper work without lingering on learning the theory. I hope that it can provide guidance and food for thought for each of you as you engage with technology and allow your digital self to evolve.

I hope you enjoy it!

Glossary

Affordances: The material properties of an environment (i.e., design of device, or study) that affects the way in which people interact with it and themselves.

Balance: A state in which elements are distributed in a manner that provides the greatest stability.

Balance Point: A state in optimal presence in which all of the states are equally present. At the optimal balance point, one has a sense of presence in which awareness of and engagement with self and other is achieved.

Devices: The different physical wearable technologies used for the purpose of the study.

Engagement: Employee commitment and passion that includes happiness, alignment, and job satisfaction.

Generativity: A concern for establishing and guiding the next generation, legacy[1]

Grokking: (ref: Stranger in a Strange Land) understanding something intuitively with one's entire being

Intention: The conscious expression of purpose in an interaction usually based on personal and cultural values and/or experience, often misunderstood based on opposing intentions when either the sender or the receiver lacks mindfulness.

Materiality: The arrangement of an artifact's physical and/or digital materials into particular forms that endure across differences in place and time[2]

Mediated Interaction: Interaction between sender and receiver that takes place through technology allowing the interaction to occur in suspended time and/or space.

Mindfulness: Presence of mind in the moment without judgment.

Optimal Presence: When the contents of extended consciousness are aligned with the other layers of the self, and attention is directed towards a currently present external world[3].

Persuasive Design: Design with the intention of promoting behavioral change.

Presence: Is a neuropsychological phenomenon; the non-mediated (pre-reflexive) perception of successfully transforming intentions in action within an external world[4].

Reciprocity: The evolutionary basis for cooperation in society[5] as the giving of benefits to another in return for benefits received (Molm, 2010).

Presenteeism: A human resources concept referring to the situation when employees are physically present but presence of mind is hindered resulting in loss of productivity. The three sub-forms of presenteeism are health-related, disengagement, and cognitive[6]

Quantum Physics: principles state that, consciousness, energy and matter are all interconnected.

Social Optimization: A social strategy term describing a method for optimizing relationships in both online and offline context. It is used to describe the methods and mindset required for building and maintaining mutually beneficial and effective relationships using social tools from events to social networks and applications.[7]

Social Strategy: Strategic application of the philosophy of social business, in which social technologies supported by new approaches facilitate a more open, engaged, and collaborative foundation for how we work[8]

Social Technologies: Digital technology that enables interaction between people: smart phones, social media, wearable tech, tablets, social gaming, augmented reality, music sharing, social shopping, social search, location-based services, and so forth.

Sociomateriality: An emerging concept referring to the fusion of social (institutions, norms, discourses, and all other social phenomena) and material (technology and other objects)[9]

Tech Nation: The concept derives from native cultures in which everything is a living being and represents something to be cared for and respected in relation to other beings.

Wearables: Electronic devices that are worn on the body, often embedded into clothing or accessories. They integrate the use of sensors and/or communications capability for connection to other devices and data.

Wellness: A state of complete physical, mental, and social wellbeing, and not merely the absence of disease or infirmity (World Health Organization, 1948).

Wellness Wearables: A category of wearable technology that is used for devices with the function related to health and/or wellness.

Workplace: Physical location where someone works. In the contemporary workplace this can be a remote home office, a café, in a large office, or any combination or variation.

GIVING BACK

To thank those who took the time to share their thoughts with me in interviews and brainstorming. I want to express my deepest gratitude for your sharing of your time and personal stories. I can't even begin to thank my proofreaders enough. I owe you big time: Eliza, Tally, Heather and Björn.

Here are links to a some that my readers might want to learn more about.

All links and references in this book are available at

http://www.2balanceu.com/digital-self-mastery-references-and-links

Jeffrey Tambor, creator of Woven Lightning, applied 15 years of deep personal transformational experience into developing a unique way of surfacing and obliterating the walls and blocks that prevent you from finding the fulfillment and success you've been searching for. The Woven lightning approach effectively upgrades your whole operating system for more leveraged and powerful results in all areas of your life. wovenlightning.com

Sheryl Bernstein **sherylbernstein.com**

Malin Olander Roese **MOOV.se**

Michelle Kopper **The Inspired Voice michellekopper.com**

Tally Forbes **Tally Forbes Gallery** https://www.facebook.com/Tally-Forbes-Gallery-135540066563105/

Simone Janssen **Salt Leadership SaltLeadership.com**

Alina Vincent **Business Success Edge businesssuccessedge.com**

About The Author

About The Author

Dr. Heidi Forbes Öste is a behavioral scientist passionate about the potential for technology and wellbeing innovations to enhance the ability to be one's best. She combines 25 years experience in social technologies and social strategy for organizations with research in presence-of-mind, wellbeing technology and the user experience. A scholar, practitioner, connector and global citizen.

Her motto: Knowledge is Power, Sharing is Powerful.

Find out more at
https://www.amazon.com/author/forbesoste

Or visit http://2BalanceU.com

Other Books By Dr. Heidi Forbes Öste
BE-ing@Work: Wearables and Presence of Mind

Upcoming editions of Digital Self Mastery coming soon:
Tech Workers edition (special edition for CES 2018)
Family - Across Generations edition (Spring 2018)

Can I Ask You For A Favor?

If you enjoyed this book, found it useful or otherwise then I'd really appreciate it if you would post a short review. I do read all the reviews personally so that I can continually write what people are wanting.

If you'd like to leave a review then please visit the link below:

http://bit.ly/2xq4Tfa

I am the first to admit that I am not perfect. In the interest of getting the Digital Self Mastery movement going, this is a first edition of many to follow. If you noticed any typos or major errors, please do not hesitate to private message me in the **Digital Self Mastery Facebook** group. I will be sure to fix it for the next edition. I will be happy to send you an updated copy in appreciation for your help.

Thanks for your support!

Footnotes

1. Erikson, E. H. *Childhood and society.* New York: Norton, 1950

2. Leonardi, Paul M. "Theoretical foundations for the study of sociomateriality." Information and Organization 23, no. 2 (2013): 59-76.

3. Waterworth, John, and Giuseppe Riva. Feeling present in the physical world and in computer-mediated environments. Springer, 2014.

4. Riva, Giuseppe. "Presence and social presence: from agency to self and others." In Proceedings of the 1th Annual International Workshop on Presence. 2008.

5. Nowak, Martin A., and Karl Sigmund. "Shrewd investments." Science 288, no. 5467 (2000): 819-820.

6. Oste, Heidi Forbes. "Be-ing@ work: Wearables and presence of mind in the workplace." PhD diss., Fielding Graduate University, 2016.

7. Öste, Heidi Forbes. "Social Optimization Theory."

8. Li, Charlene, and Brian Solis. The seven success factors of social business strategy. John Wiley & Sons, 2013.

[9a] Orlikowski, Wanda J., and Susan V. Scott. "10 sociomateriality: challenging the separation of technology, work and organization." Academy of Management Annals 2, no. 1 (2008): 433-474.

[9b] Leonardi, Paul M. "Theoretical foundations for the study of sociomateriality." Information and Organization 23, no. 2 (2013): 59-76.

[9c] Parmiggiani, Elena, and Marius Mikalsen. "The facets of sociomateriality: A systematic mapping of emerging concepts and definitions." In Scandinavian Conference on Information Systems, pp. 87-103. Springer, Berlin, Heidelberg, 2013.